EARLY BIRD STORIES™

Let's Notice Types of Materials

Martha E. H. Rustad   Illustrated by **Christine M. Schneider**

LERNER PUBLICATIONS ◆ MINNEAPOLIS

NOTE TO EDUCATORS

Find text recall questions at the end of each chapter. Critical-thinking and text feature questions are available on page 23. These help young readers learn to think critically about the topic by using the text, text features, and illustrations.

Lerner Publications Company
An imprint of Lerner Publishing Group, Inc.
241 First Avenue North
Minneapolis, MN 55401 USA

For reading levels and more information, look up this title at www.lernerbooks.com.

Photos on p. 22 used with permission of: iamlukyeee/Shutterstock.com (girl playing with sand); corbac40/Shutterstock.com (states of matter diagram); Gabriel Pahontu/Shutterstock.com (glassblower).

Main body text set in Billy Infant.
Typeface provided by SparkyType.

Library of Congress Cataloging-in-Publication Data

Names: Rustad, Martha E. H. (Martha Elizabeth Hillman), 1975- author. | Schneider, Christine M., 1971- illustrator.
Title: Let's notice types of materials / Martha E. H. Rustad ; illustrated by Christine M. Schneider.
Description: Minneapolis : Lerner Publications, [2022] | Series: Let's make observations (Early bird stories) | Includes bibliographical references and index. | Audience: Ages 5-8. | Audience: Grades K-1. | Summary: "Ms. Sampson's class goes on a treasure hunt to see how different materials look and feel. Young readers will follow along and think about why items are made of certain materials"—Provided by publisher.
Identifiers: LCCN 2021012236 (print) | LCCN 2021012237 (ebook) | ISBN 9781728441368 (lib. bdg.) | ISBN 9781728444680 (eb pbf)
Subjects: LCSH: Materials science—Juvenile literature. | Matter—Properties—Juvenile literature.
Classification: LCC TA403.2 .R869 2022 (print) | LCC TA403.2 (ebook) | DDC 620.1/1—dc23

LC record available at https://lccn.loc.gov/2021012236
LC ebook record available at https://lccn.loc.gov/2021012237

Manufactured in the United States of America
1-49907-49750-5/17/2021

TABLE OF CONTENTS

CHAPTER 1

A NEW PROJECT

Our teacher, Ms. Sampson, puts something in a bag. Thea reaches in and tells us what it feels like. Our job is to guess what it is.

"It's round and **hard**," Thea says. "It's a flat circle and a little **bumpy**."

Jae guesses a quarter. She's right!

Coins are made from **metal**, a type of material. Materials are what we use to make things.

We are going to learn about materials in our new project.

✓Check! What are materials?

We look for materials in our classroom. We find metal, **glass**, and more.

- Metal
- Rubber
- Wood
- Glass
- Plastic
- Fabric
- Rock

Ms. Sampson

Ms. Sampson says people choose different materials for different purposes. "Why can't you make a window out of **wood**? Or a shirt out of glass?"

"Windows have to be clear!" Amal says.

"Shirts have to be soft," Olivia adds.

Then we describe materials. We say how they feel and look.

The spoon is hard and smooth. The sand is **soft**. The coins are **shiny**.

- Hard or Soft
- Bendy or Stiff
  Smooth or Rough
  Light or Heavy
  Colors
  Clear or Not
  Shiny or Dull

✓Check! Why can't shirts be made of glass?

CHAPTER 3
TREASURE HUNT!

We go on a treasure hunt around the school.

We search for different materials.

Desks, gym floors and doors, and craft sticks are made of wood.

Paper clips, coins, lockers, and doorknobs are made of metal.

We also search for **rock**, **plastic**, glass, and other materials.

Back in our classroom, we talk about what we found.

	Hard or Soft	Bendy or Stiff	Smooth or Rough	Light or Heavy	Colors	Clear or Not	S
Metal	hard	stiff	smooth	both	gold, silver, copper, and more	not clear	
Rubber	soft	bendy	smooth	light	pink, blue, yellow, and many more	not clear	
Wood	hard	stiff	rough	light	mostly brown	not clear	
Glass	hard	stiff	smooth				
Plastic	hard and soft	both	smooth	light	many colors	both	sh
Fabric	soft	bendy	rough or smooth	light	many colors	not clear	
Rock	hard	stiff	smooth	heavy	grey, brown, black, and more	not clear	

"Both wood and metal are hard," Izzy says.

"Fabric and rubber are **bendy**," Jae adds.

We learned about a lot of materials.

Time to make a sculpture in art class!

LEARN ABOUT TYPES OF MATERIALS

There are many types of materials. Materials can be hard or soft, stiff or bendy, shiny or dull, or rough or smooth.

Sand is made of tiny pieces of hard rock. Sand feels soft because wind and water have worn down the rocks and made them smooth.

Some materials change with temperature change. Metal and glass soften when heated. Then people can form them into different shapes.

All materials are made of matter. Matter is anything that takes up space. The three forms of matter are solid, liquid, and gas.

Natural rubber comes from rubber trees. But many rubber objects are made with human-made rubber.

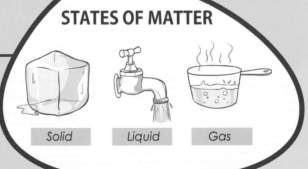

STATES OF MATTER

Solid   Liquid   Gas

THINK ABOUT
TYPES OF MATERIALS:
CRITICAL-THINKING AND
TEXT FEATURE QUESTIONS

**What materials can you
find in your home?**

**How would you describe
a coin's look and feel?**

**Where is the table
of contents?**

**How many chapters are
in this book?**

LERNER
SOURCE™

Expand learning beyond the printed book. Download free, complementary
educational resources for this book from our website, www.lernerresource.com.

GLOSSARY

describe: to tell or write about something

material: a substance people use to make things

metal: a hard material that mostly comes from rock in the ground

purpose: a reason or plan

rubber: a bendy, stretchy material

LEARN MORE

Bailey, Jacqui. *Let's Investigate Materials.* New York: Crabtree, 2021.

Kington, Emily. *Materials: What Is Stuff Made From?* Minneapolis: Hungry Tomato, 2020.

Materials
https://kids.britannica.com/kids/article/materials/476293

INDEX